MINIBEASTS UP CLOSE

Pill Bugs
Up Close

Greg Pyers

 Raintree

Chicago, Illinois

© 2005 Raintree

Published by Raintree, a division of Reed Elsevier, Inc.

Chicago, Illinois

Customer Service 888-363-4266

Visit our website at www.raintreelibrary.com

For information, address the publisher:
Raintree, 100 N. LaSalle, Suite 1200, Chicago, IL 60602

09 08 07 06 05
10 9 8 7 6 5 4 3 2 1

Printed and bound in Hong Kong and China by South China Printing Company Ltd.

Library of Congress Cataloging-in-Publication Data

Pyers, Greg.
  Pill bugs up close / Greg Pyers.
      p. cm. -- (Minibeasts up close)
  Includes bibliographical references (p.   ).
  ISBN 1-4109-1531-X (hc) -- ISBN 1-4109-1538-7 (pb)
  1. Isopoda--Juvenile literature. I. Title. II. Series.
  QL444.M34P94 2005
  595.3'72--dc22
                                    2004022291

Acknowledgments
The publisher would like to thank the following for permission to reproduce photographs:
© Steve Hopkin/ardea.com: pp. **12–13, 19, 24**; © Dwight Kuhn: pp. **6, 22, 23, 27, 29**; © Naturepl.com/ Niall Benvie: p. **10**, /© Dan Burton: p. **7**, /© Duncan McEwan: p. **26**; Lochman Transparencies/Dennis Sarson: pp. **8, 18**, /Jiri Lochman: p. **14**, /Peter Marsack: pp. **4, 28**; photolibrary.com: pp. **15, 16**, /SPL: p. **11**; © Paul Zborowski: pp. **5, 25**.

Cover photograph of a pill bug reproduced with the permission of Naturepl.com/Niall Benvie.

Every effort has been made to contact copyright holders of any material reproduced in this book. Any omissions will be rectified in subsequent printings if notice is given to the publisher.

Contents

Any words appearing in bold, **like this**, are explained in the Glossary.

Amazing Pill Bugs!

Have you ever seen pill bugs? Pill bugs are flat creatures that scurry in dark, damp places. You may have seen many pill bugs in a compost pile, among dead leaves, or under a pile of wood. Perhaps you have seen one curl up into a ball, though not all do this. When you look at them up close, pill bugs really are amazing animals.

Pill bugs are often found in large groups under logs.

There are more than 3,500 kinds, or **species,** of pill bugs.

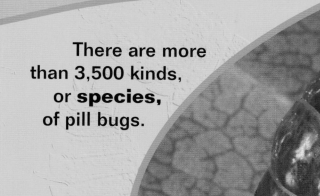

Other names

Pill bugs are also known as slaters, sow bugs, and wood lice. Some people call a pill bug a "roly-poly."

What are pill bugs?

Pill bugs are crustaceans. Crabs, crayfish, and barnacles are also crustaceans. Crustaceans have no bones. Instead, they have a hard, tough skin, called an **exoskeleton.** Crustaceans have many legs and most kinds live in water. Unlike most crustaceans, pill bugs live on land.

Where Do Pill Bugs Live?

Pill bugs are found in many different parts of the world. They live in the hottest deserts of Africa and in salty pools in Australia. The common sea slater lives on beaches. Most pill bugs live in forests.

Habitat

A **habitat** is a place where an animal lives. Pill bugs are found in a lot of different habitats. Most live in damp, dark places. The **leaf litter** on a forest floor is a good habitat for pill bugs. Pill bugs are also found in compost, soil, and under bark and rocks.

The cracks in brick walls can be home to many pill bugs.

Pill bugs that burrow

Desert pill bugs dig burrows to shelter in during the hot day.

Pill bugs live in these places because they find their food there. They can also stay hidden from **predators.**

Many pill bugs move out into the open at night to feed. During the day, beach pill bugs find shelter under rocks, seaweed, and driftwood. They come out to feed when the sun has set.

Living in dark, damp places keeps pill bugs from drying out.

Pill Bug Body Parts

A pill bug's body has three main parts. These are the head, the **thorax,** and the **abdomen.** The body is covered by a hard **exoskeleton.**

The head

A pill bug's head has a mouth, two eyes, and two pairs of feelers called **antennae.** One pair is very small and difficult to see.

The thorax

In adult pill bugs, the thorax has seven parts, or **segments.** Each segment has a back plate. These look like pieces of **armor.** The pill bug's seven pairs of legs are attached to these segments. A pair of legs is attached underneath each segment.

The abdomen

The abdomen is much shorter than the thorax. At the end of the abdomen there are two tail-like body parts called **uropods**.

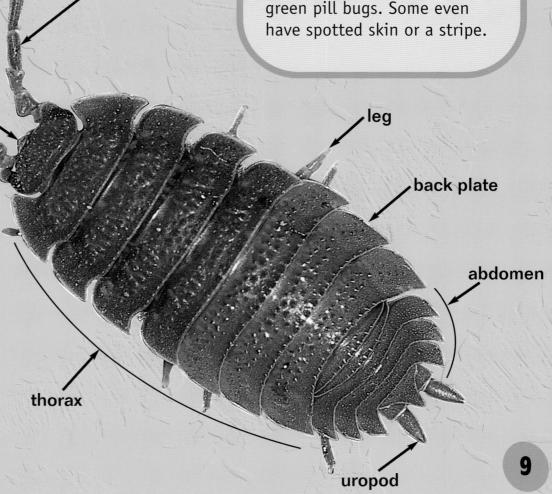

antenna

head

leg

back plate

abdomen

thorax

uropod

Pill bug colors

Many pill bugs are dark gray in color. But there are also red, orange, brown, cream, and green pill bugs. Some even have spotted skin or a stripe.

Mouthparts and Eating

Most pill bugs eat rotting plants. In a compost pile, pill bugs eat potato peelings, cabbage leaves, tomatoes, and carrot tops. Pill bugs also eat **fungi** that grow on leaves. Quite often, they eat their own droppings.

Pill bugs sometimes eat the flesh of dead animals. They may eat other pill bugs, even when they are alive. This may happen when a pill bug is shedding its skin. At that time, its soft body is easy to bite.

On a forest floor, pill bugs eat dead leaves and wood.

Mouth

The mouth has two jaws, called **mandibles.** These break food into small pieces for swallowing.

Drinking

Pill bugs get water in several ways. One way is from the moist food they eat. Another is to drink it, for example, from dewdrops. A third way a pill bug gets water is by taking it in through its **uropods.**

A bee—eating pill bug

The sand beach pill bug of New Zealand mainly eats the bodies of honeybees that have drowned on the seashore.

A pill bug's mouth is on the bottom of its head.

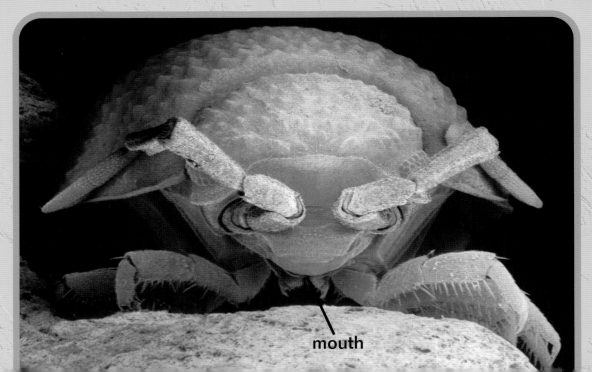

mouth

11

Droppings and Nutrients

Pill bugs recycle their waste by eating their droppings.

Why do pill bugs eat droppings?

When a pill bug swallows food, the food moves through a long food tube to the stomach. As it moves along, the food is broken down. This releases **nutrients** that the pill bug needs to stay alive. The nutrients are taken into the pill bug's blood. Some nutrients may pass out through the anus in the pill bug's droppings. By eating the droppings, a pill bug can obtain these nutrients.

Copper

One nutrient a pill bug must have is copper. Copper is a metal, such as iron or aluminum. The copper in the pill bug's blood carries **oxygen.** Pill bugs get copper by eating rotting leaves. When there are no rotting leaves around, a pill bug can get copper from its droppings.

Blue blood

In human blood, it is iron that carries oxygen. Iron gives our blood a red color. In pill bugs, copper gives blood a blue color.

Rotting leaves supply copper for this pill bug.

13

Seeing and Sensing

Pill bugs **sense** the world around them in several ways.

Eyes

A pill bug has **compound eyes.** Each compound eye is made up of many very small eyes. Each small eye faces in a slightly different direction. It sees something a little bit different from the other eyes.

Some insects' compound eyes have thousands of small eyes. A pill bug's compound eyes have just fifteen to twenty small eyes. This means that a pill bug's eyesight is poor.

compound eye

A pill bug does not need good eyesight because it lives in dark places.

Antennae

A pill bug has two pairs of feelers called **antennae.** One pair is large. As the pill bug walks, it taps these antennae on the ground in front of it. The antennae pick up smells and enable the pill bug to find food. Smells can also lead a male pill bug to a female.

The second antennae, called **antennules,** are tiny and probably have no use.

Folding antennae

When a pill bug rolls into a ball, it neatly folds away its antennae into hollows in its head.

A pill bug uses its large antennae to find food.

antenna

15

Legs and Moving

Pill bugs have fourteen legs. The legs are arranged in seven pairs. Each pair is attached underneath a pill bug's body. There is a pair of legs attached to each of the seven **segments** of the **thorax.**

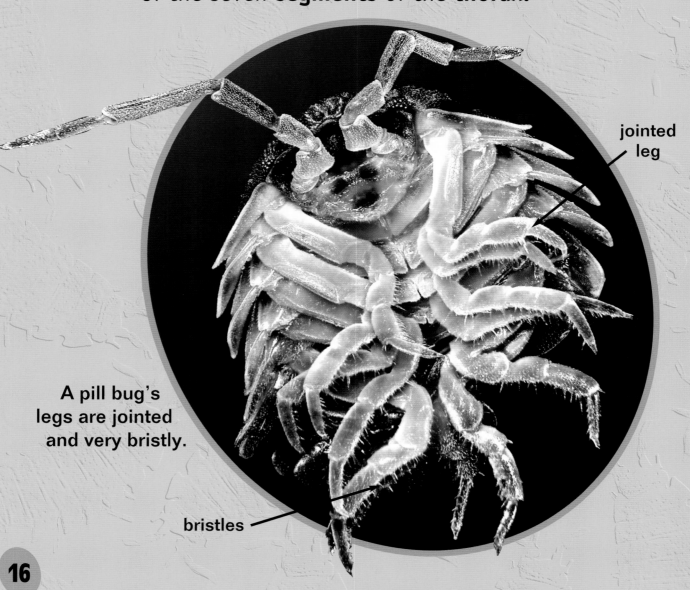

jointed leg

A pill bug's legs are jointed and very bristly.

bristles

Legs

Pill bug legs are jointed. This means that they have separate sections joined together. All fourteen legs are the same size and shape.

Walking and running

Pill bugs can move at different speeds. When they are searching for food, they move slowly. When a light comes on, or when a **predator** is about, they scurry for cover.

A pill bug does not walk just on its feet. Half of each leg also touches the ground. The lower part of each leg has short bristles. These give the pill bug a good grip on leaves, rocks, and sticks as it walks.

Protecting Themselves

Many animals eat pill bugs. There is a **species** of spider that eats nothing but pill bugs. Centipedes, beetles, frogs, and newts eat pill bugs. In Europe, hedgehogs and shrews eat pill bugs.

Avoiding predators

One way that a pill bug avoids **predators** is to hide. Some pill bugs scurry away when danger threatens. Others can roll into a ball. This protects their soft undersides from attack by small predators, such as centipedes. Rolling up may also confuse a predator.

Pill bugs can give off an unpleasant smell. This keeps many predators away.

Some pill bugs, such as this one, can roll into a ball. The other pill bug pictured cannot roll into a ball.

Drying out

Many species of pill bugs lose water quickly through their **exoskeletons.** Their exoskeletons are not waterproof. But this is not a problem as long as these pill bugs remain in damp places.

Desert pill bugs avoid losing water by staying in their burrows by day. They come out to feed after dark.

Inside a Pill Bug

The inside of a pill bug is a lot like the inside of an insect.

Blood and heart

A pill bug's blood moves through the spaces in its body. The heart is long and tube-shaped. It runs beneath the **exoskeleton** along the pill bug's back. Blood travels from the head, through the **thorax** and **abdomen,** and then the heart pumps it forward again.

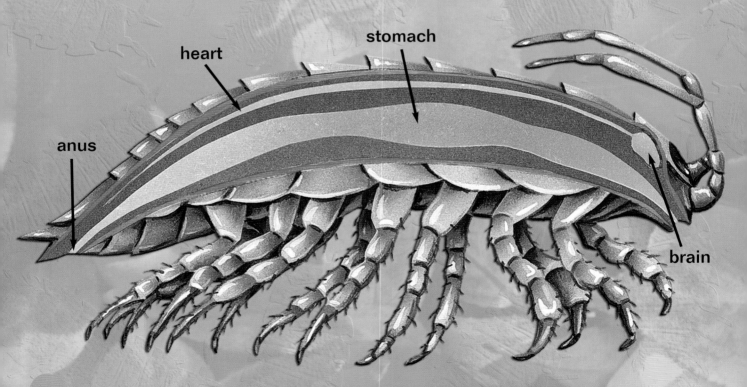

heart

stomach

anus

brain

How do pill bugs get air?

Pill bugs do not breathe air in through their mouths. A few **species** take in **oxygen** through groups of tiny tubes underneath the back end of their bodies. Even if the pill bugs have the tubes, most take in oxygen through **gills** under the abdomen. Inside the body, oxygen is taken into the pill bug's blood.

The brain

A pill bug's brain gets information that it **senses** through its **antennae** and eyes. It sends messages to the rest of the body about what to do.

Pill Bug Eggs

Pill bugs hatch from eggs in a pouch under the mother's belly. The eggs start to develop there after she **mates** with a male pill bug.

Life in the pouch

Dozens of eggs may hatch inside a pill bug's pouch. But the babies do not leave the pouch. They stay there to grow and develop. The pouch is full of **fluid.** This stops the babies from drying out.

Pill bug eggs do not have shells.

This pill bug is carrying eggs in her pouch.

The pouch is above the first five pairs of the mother's legs. The top halves of these legs lie quite flat against the pouch. They protect the pouch from damage if the mother walks on rough ground or rolls into a ball.

Leaving the pouch

After one to three months, the young pill bugs are ready to leave the pouch.

Producing alone

The females of some pill bug **species** produce young without mating with a male.

Leaving the Pouch

At first, baby pill bugs look very much like adult pill bugs. There are a few differences. They are much smaller and much lighter in color. This is because their **exoskeletons** have not yet hardened. Young pill bugs have only six pairs of legs.

These young pill bugs stay rolled up to protect themselves from **predators,** such as centipedes.

24

Growing

At this early age, young pill bugs are called **mancas.** Within a day of leaving its mother's pouch, a manca has grown too big for its exoskeleton. It **molts,** which means that the old skin splits and the pill bug crawls out with a new, bigger exoskeleton. After the second molt, the pill bug is called a **juvenile.**

Pill bug life span

Some pill bugs may live for four years. Most live for just two years.

A juvenile pill bug has seven pairs of legs.

25

Getting Bigger

As a pill bug grows to adult size, it **molts** several times.

Molting

Molting takes place in two stages. First, the **exoskeleton** covering the rear of the pill bug loosens. This makes one half of its body paler than the other half. The exoskeleton splits and the pill bug pulls itself free.

This pill bug has just molted the front half of its exoskeleton.

The second stage follows a few days later. The exoskeleton covering the head and first half of the pill bug's body falls off.

Dangerous time

Molting is a dangerous time for pill bugs. This is because the new exoskeleton is soft. Until it hardens, it provides little protection against a **predator's** jaws.

A molting pill bug may be attacked by several other pill bugs at the same time.

This pill bug will now eat its old exoskeleton. It uses the **nutrients** in the old exoskeleton to grow a new skin.

Pill Bugs and Us

Pill bugs cannot bite or sting. They do not carry diseases that make people sick. They do not harm people at all. They break down dead plants and compost to make **fertile** soil. But still, many people do not like pill bugs. Why?

Maybe it is because pill bugs have many legs and remind us of spiders. Perhaps it is because pill bugs are seen in dark, damp places where there are rotting plants. These are the kinds of places people do not like.

Pill bugs do an important job in compost piles, breaking down food scraps.

Pill bugs make good pets

Some people keep pill bugs as pets. A plastic or glass container with some soil, dead leaves, and a few twigs makes an excellent pill bug **habitat.** A spray of water now and then will keep it moist. The pill bug can be given a few pieces of fruit or a vegetable peel to eat.

You can learn a lot about pill bugs by keeping them in a jar or box.

Find Out for Yourself

You may be able to find some pill bugs in a yard. Look in the soil, among the **leaf litter,** and under rocks and logs. You could keep some pill bugs in a container and watch them up close.

Books to read

Murray, Peter. *Mollusks and Crustaceans*. Eden Prairie, Minn.: Child's World, 2004.

Pascoe, Elaine. *Pill Bugs and Sow Bugs: And Other Crustaceans*. Farmington Hills, Mich.: Gale Group, 2001.

St. Pierre, Stephanie. *Pillbug*. Minneapolis, Minn.: Sagebrush Education Resources, 2002.

Using the Internet

Explore the Internet to find out more about pill bugs. Have an adult help you use a search engine. Type in keywords such as *pill bug, wood lice, wood louse, sow bug,* or *slater*.

Glossary

abdomen   last section of a pill bug's body

antenna   (more than one are called antennae) feeler on a pill bug's head

antennule   small antenna of a pill bug

armor   hard covering that protects the body

compound eye   eye made up of many small parts

exoskeleton   hard outside skin of a pill bug

fertile   having many nutrients

fluid   something that is runny, not hard, such as juice

fungus   (more than one are called fungi) plantlike living thing that feeds on dead plants and animals

gills   body part of some pill bug species that allows them to breathe

habitat   place where an animal lives

juvenile   young

leaf litter   dead and rotting leaves on the forest floor

manca   young pill bug before it molts for the first time

mandible   jaw

mate   when a male and a female come together to produce young

molt   when a growing pill bug splits open its outside skin and climbs out of it

nutrients   parts of food that are important for an animal's health

oxygen   gas in the air that is needed for life

predator   animal that kills and eats other animals

segment   one of the separate parts of a pill bug's body

sense   how an animal knows what is going on around it, such as by seeing, hearing, or smelling

species   type or kind of animal

thorax   chest part of a pill bug

uropod   short tail at the rear end of a pill bug

Index